Logbook Story

Dr Ganesh Janardan Ghugare

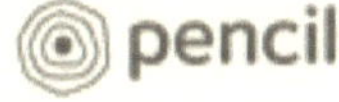

pencil

ISBN 978-93-5667-404-2
© Dr Ganesh Janardan Ghugare 2023
Published in India 2023 by Pencil

A brand of
One Point Six Technologies Pvt. Ltd.
123, Building J2, Shram Seva Premises,
Wadala Truck Terminal, Wadala (E)
Mumbai 400037, Maharashtra, INDIA
E connect@thepencilapp.com
W www.thepencilapp.com

Author biography

Dr. Ganesh Janardan Ghugare
Entrepreneur (Education & Entertainment)
Film Maker – Director - Producer (Feature, Web-Series, Theatre)
Author (Worldwide, English, Marathi)
Artist (Actor, Drawing & Painting)

EDUCATION
MA- Master of Arts in History
ATD- Art Teachers Diploma
AM- Art Masters in Drawing and Painting
PGDB- PG Diploma History
LL.B.- Bachelor of Law
PH.D- In History
MDNMS - Diploma Course in "Mastering Direct & Network Marketing Skills & Concepts"

CONTACT
B- 410 NITYANAND KRUPA , RAM GANESH GADKARI ROAD , PRABHU ALI, OLD PANVEL, PANVEL, MAHARASHTRA -410206.
EMAIL: drganeshghugare@gmail.com   ,
ghugaregj@yahoo.co.in

FIELD EXPERIENCE

As a freelance trainer and speaker 15 years (Education, Art, Law, Films, Management)

Educational campaigning with Edubridge India platform throughout Maharashtra 2016 to till date (Competitive Examinations and Olympiads).

As an author, written and published, 7 Books on history and art, Worldwide

Students to Artist, (English) Art : 2014,
Chh. Shivaji Complete History 1630-1680 AD, (English) History : 2020,
Military Design Of Chhatrapati Shivaji, (English) History : 2020 ,
Life Graph of Chh. Shivaji Maharaj, (English) History : 2020,
Unbelievable battles moves of Chh. Shivaji, (English) History : 2020,
Dishaheen (Marathi) Poetry: 2021
A Closed Volume, (English) Short Stories : 2022

As an Artist 22 Paintings Exhibitions : Mumbai, Navi Mumbai, Panvel and Pune

As a Film Director : film makers many (Nos.26) short films and web-series

Theatre, Film and Web Series.

Fayaan Marathi Drama At Damodar Theatre, Parel, Mumbai : 2022
Mumbai Carpet : Panvel : 2016

Aarambh (Marathi) : Webseries 2018
Aarambh (Hindi) Webseries on M2M Studio OTT : 2021
Zunjar Marathi Feature Film on ABC Talkies OTT : 2022

AWARDS AND APPRECIATIONS
Total 18 awards for achievement and contributions in school education sector. Attended many CBSE sessions and international summit.
Research paper published in ISSN magazines.
Participated national/ international level research conferences.
PERSONAL
Date of Birth   :   14th September, 1985
Nationality   :   Indian
Marital Status   :   Married
Languages Known   :   English, Hindi and Marathi.

CONTENTS

Introduction

Logbook story is written by
Dr Ganesh Janardan Ghugare

Short stories included:
Humanity Underground.............
Ahiravan ....
Ghera...................................
Dandak..................................
Question Mark........................

Story No. 01. Humanity Underground

This story is very old. Raipatan taluka in Maharashtra was a topic of discussion in Maharashtra. The incident was in 1993. A settlement in Raipatan was inhabited by Hindu-Muslims as well as people of various religions and sects. People are fighting over caste and religion. It would go on but the feud never led to any genocide. After the demolition of Babri Masjid, the rift between the Hindus and Muslims had repercussions in Patan and it was not surprising, but the communal riots that took place at this time were very terrible. A massacre like this has never happened anywhere in India, a massacre like this happened in Raipatan. There was a lot of bloodshed between Hindus and Muslims.

Each other's houses were burnt. Nanny-sisters were robbed of their dignity. Children were also slaughtered. Wild forest burns, and so on, so humanity has become in this riot. To tell the truth, this murder was done by humanity. That story is from the same time. Fearing the carnage going on in Raipatan, a man took his two children and entered an abandoned house. He was terrified. He wanted to save his children and his life.

Seeing the situation in Raipatan, military forces came from Delhi and captured Raipatan. A kind of curfew was imposed here. No one was allowed to leave the house. If anyone was seen, he was ordered to be shot on the spot immediately. The children's father entered the house and took the children and was looking outside in a panic as there were sounds of commotion outside. He was very scared when the bullets were firing. His 10-year-old son and 6-year-old daughter were both scared and sitting next to him. Suddenly he noticed that something was hurt on his back near his waist. He realized that he had been shot and was bleeding profusely.

This man, who is a Hindu by religion, was praying to God to save us from this crisis, but at this time, God was also stunned. The children did not know who to ask for help. All they knew was that their father was injured. They knew his pain but their age and understanding were too small, so they could not help their father in any way and if he sought help from anyone outside, he would surely die, either by Muslim assailants fuelled by communalism or by a shoot-at-sight. From jawans who received orders. But death was inevitable upon exit. The father was desperately trying to stop his bleeding. He was also trying to remove the bullet from the body. Finally the boys followed his instructions and removed the bullet from his body, but the bleeding was not ready to stop as time passed.

There was a noise from outside, jawans were loudly announcing shoot at sight orders if anyone was seen on the street. After some time, the rioters could also hear their voices. Sometimes he said Jai Shriram, sometimes Allahu Akbar. The children were hungry. The father searched the house and gave the children something to eat. Now his vision started getting blurry. Then he realizes that he has few hours left of his life.

Even though the children saw him, the bleeding would not stop. He was badly injured. Looking at his two childrens, he remembered his happy days. The sudden attack on the girl's birthday celebration in a brutally communalized Muslim settlement, his escape, the runaway Muslim mob, his deadly pursuit, the bullet wound to his wife and his escape with his children leaving her dead body there. . He started remembering all these things. He took both the children and started crying. What about my children after me? He began to face this question, and within it he was lost in a void.

His son understood that his father was in ·pain now. And….. something would happen to him the girl was sick. Slowly looking at the children, even the poor man did not know when the father had left his life. These two children were holding hands and could go out. They don't know how to cope with the situation they were in.

The boy came to know that his father had left this world, but the girl had no idea. After the father's death, the ground shifted from under the son's feet, and momentarily and momentarily, he took his father's place to protect his sister. He was aware that his death was inevitable when he went out, but he would die even if he stayed here because there was nothing to live, eat, live in that locked room. A day passed and his father's body started to rot. His smell began to spread. Both the children's started vomiting from the smell. Both the children's were staring at their father's dead body.

They would eat whatever they could find in the hiding place... While both the childrens were sleeping, suddenly, someone entered the room by opening the door and closed the door and the man went and sat in the corner. This man was injured. He was shot in the hand. This man, who was a Muslim, was looking at both the children for a long time. He was looking at the rotting corpse of their father. Also these two scared children were staring at the man, thinking that he is going to kill us. The Muslim man remembered the tragedy of his family. He remembered the slaughter of his wife and son and the bullet that hit him during the curfew and from there he came to the secret place while reading. He remembered the incident and started taking out his anger on both the kids. He started beating both the kids. His daughter's face came in front of him as the girl's throat was cut. Yes, he started crying like an ox, both the children were scared and time was passing.

Both children were hungry. Both children were now eating banana peels. The Muslim man was eating the food he had. Seeing him eating, these two little kids were staring at him. Seeing these two children looking at him, the Muslim man offered them some food too. The kids were a little scared at first, but then they started eating it.

The boys got some patience now the boy helped him to remove the bullet from his hand. It was he, who removed the bullet from his father's waist. He reminded the Muslim man of this with the help of the boy, the Muslim man wrapped the dead body of the boy's father in a plastic and kept it aside. The Muslim man saw his children in those two little boys. Both the children's also now developed an affection for the Muslim man as suddenly they heard the announcements of raising the curfew, the Muslim man became alarmed.

The military jawan broke the door and army jawan came in with his hand and pointed his gun at this muslim man and asked him are you Hindu..or Muslim..? After being silent for a while the Muslim man used only one word and uttered it.... Man! Whose children are these? The Muslim man immediately replied without getting stuck that both the children are mine. Caste religion separates man only humanity in man unites man.

Story No. 02 Ahiravan

Samayara Rajput is talking with a man.... very tired The man is sitting in front of her. The room is not cleaned, the stuff is scattered, the room is like a chamber. Only the back of the man is visible. Samayara is narrating her story to him.

Dark night.... In this dark night of heavy rain, a car accident happens. It's raining After the accident of that car, a man's eyes are visible there. His teeth are visible when he smiles. And he laughs out loud. One of his teeth is made of gold. The name of this man is Shravan Bhogi.

Samayara Rajput, Age 30 Years, Beautiful, Well Educated...Doctor in City Private Hospital..Samayara Rajput dreaming in her sleep...Suddenly she wakes up, very scared. Samayara Rajput.... Sitting on the bed lifts the bowed neck.

Sitting on the bed, looking in front of him, she sees.... Shravan Bhogi shouts and says.

Shravan Bhogi :

"will not leave

won't leave you

I will destroy everything related to you

I will erase your personality"

Gets out of bed at the same time. Walks while thinking Is upset Samayara then sits on the bed. Little by little she is afraid.

Samayara Rajput gets thinking "Who was that? Why was he saying like this? What did I do to him? I didn't know?"

For the last one year, he used to appear in Samayara's sleep. That's why she had not slept for the last one year. Neither Samayara had ever seen him, nor had she met him, yet Samayara felt so. This face is familiar. Samayara remains engrossed in those dreams throughout the night. Samayara wakes up in the morning, gets ready and goes to the hospital. Samayara Rajput is a doctor (obstetrician), pregnancy and child specialist. She remains busy in the hospital throughout the day. She comes home from hospital work and goes to sleep.

A year ago, Samayara Rajput had a big accident.

In that accident, Samayara's friend, the administrator of the hospital Dr. Raunak Verma and Dr. Samayara's parents died. This incident happened at night while returning from the hospital party. Then Dr. Samayara was in sleep.

Raunak was a very good doctor and also a friend of Dr. Samayara. He was also on the board of directors of the hospital.

Along with the treatment of all the special cases of Hospital, she also used to do counseling for pregnant women and children.

Dr.Samayara goes to her hospital. There she sits on a chair and drinks water kept on the table. And later calls her receptionist Mary. Dr. Samayara's assistant Vicky D'Souza did not come even today and kept the phone switched off. His girlfriend Jennifer...Janny had been a student of Dr. Samayara.

Somewhere far away….. Shravan Bhogi catches a woman and recites a poem. Shravan Bhogi caresses the hair of that woman. Shravan Bhogi killed the woman while reciting poetry. Her baby was taken out by cutting her stomach. Removes the heart from the chest of a woman. Shravan Bhogi laughs. In front of a frightening idol of a goddess, he cuts the infant taken out of a woman's stomach into pieces and offers it to the goddess. A woman's heart eats considering it as Prasad.

Dr.Samayara reminisces about her past while drinking coffee in her balcony. Shravan Bhogi

comes in a dream and says that

Shravan Bhogi :

"will not leave

won't leave you

I will delete everything related to you

I will erase your personality"

While drinking coffee, Samayara remembers something. She picks up the phone and dials the number, the ring keeps ringing in front of her and the phone gets disconnected. Samayara gets disturbed. Tries the phone once again. Then the ring rings in front and Vicky picks up the phone.

Vicky tells her that he had a minor accident. Because of that he cannot come to the hospital for the next 15 days. Samayara tells him about the unknown man seen in her dreams. Vicky tells her that it is all Samayara's illusion.

Vicky advises Samayara, if she is not sure. So she should talk to all the relatives associated with her, and check whether they are all right or not? Samayara talks to all her relatives. Everything is fine. Samayara is convinced that Vicky was right. This is all Samayara's illusion.

This dream is just a dream… nothing is an accident……it is going to happen… thinking of living life positively, Samayara finalizes going to the hospital tomorrow.

Somewhere far away…. Shravan Bhogi is carrying a woman while dragging her to a deserted place. That woman screams a lot, but Shravan pulls out Arbhaka from her stomach, the woman slowly calms down and dies. Shravan Bhogi kills that woman like this by ripping her stomach. And laughs out loud. Shravan Bhogi removes the heart of this woman too.

The next day, Dr. Samayara Rajput goes to the hospital. The appointment of her patient Mrs. Sengupta remains fixed today. But as she does not turn up on time, she tells her assistant cum receptionist, Mary, to call her. Then

Mary tells Samayara. That Mrs. Sengupta is dead. Dr. Samayara Rajput gets shocked. She calls on Mrs. Sengupta's number, then her brother-in-law Alok Sengupta picks up the phone and tells the whole thing. Mrs. Sengupta, a patient of Dr. Samayara Rajput was abducted by someone and a complaint was lodged at the police station. After 2 days, the dead body of Mrs. Sengupta was found lying on the road by the side of a forest. Someone had taken away her 8-month-old infant by cutting her stomach.

Dr. Samayara Rajput gets stunned after hearing this. Because the same incident happened with Mrs. Salunkhe a few months back. Samayara Rajput calls Vicky and tells that this incident has happened with Mrs. Sengupta. Vicky also gets upset because he also knew that a few months back this year a similar incident had happened with Mrs. Salunkhe. Samayara Rajput tells Vicky that the murderer of both of them would be the same.

On the other hand... Somewhere far away, Shravan Bhogi sacrifices the infant brought from the womb of Mrs. Sengupta in front of the goddess. breaks it into pieces. The idol of this goddess is different.

Vicky D'Souza and Jenny both come to the hospital at the call of Samayara Rajput. Explore old data. Then they come to know that, Total 6 patients had suddenly stopped coming to the hospital. This calls Vicky D'Souza to 6 clients. Then it is known that 2 of these murders have happened like this. And the remaining 4 are still missing. The dead bodies of two of these patients were found. He was also murdered like this. All the women were in their eighth month. The murderer had taken away her child after cutting her stomach and had taken out her heart.

As Samayara Rajput comes to know about this. So she calls Inspector Pandurang Bagdi, who was handling Mrs. Sengupta's case. Pandurang tells Bagdi that "Like Mrs. Sengupta, 2 months ago, patient Mrs. Salunkhe was also murdered in the same way… in the village. His village was in Gadchiroli and I got suspicious because in the last few months my patients suddenly disappeared like this. Those who stopped coming. When I inquired with them, I came to know that two more of them were murdered. He was in the village and the remaining 4 are still missing. All these four murders are of my patients. I feel like this case is related to me.

Samayara Rajput tells Vicky that, "Maybe this man in dreams was talking about erasing anything related to me, isn't he the murderer of these women?"

Pandurang Bagdi calls Samayara and tells her. "If this murder mystery is related to you, then the data of pregnant women has been leaked from your place and 3 months back. Because all of them were admitted to your hospital 3 months back, so it is confirmed that your data has been leaked.

Samayara Rajput requests Pandurang Bagdi to keep her in the team with the police in this investigation. Inspector Pandurang Bagdi denies this. On the contrary, he says, "You are also under suspicion" because all of them were Samayara's patients.

Samayara, Vimal, who was Samayara's receptionist in her hospital 3 months back. She calls her. So her phone switches off. Then after finding her left diary in the hospital, she gets her neighbor's number. After calling, Samayara Rajput on the neighbor's phone, the neighbor tells her that she left the village 3 months ago and we do not have her husband's number. Because, her husband is a very bad man. He looks at the women of the neighborhood with dirty eyes.

Kamal's number is received from the neighbor of Vimal's sister. Samayara Rajput calls Kamal. After calling, Kamal tells her that Vimal died 3 months ago. She had gone to the village with her husband. Vimal's husband told kamal that she was bitten by a snake and she died. Burned it too. We were not even called. Her husband Shravan Bhogi is a very dirty man and also teases women.

Kamal tells that, "Last year ago he had an accident with a doctor. Shravan had hit the doctor's car after drinking alcohol. After that he was not fit to give birth to children. There was no fault of that doctor in that. But due to this rascal, the doctor's car met with an accident. That poor doctor died. He had a doctor friend inside the car and there were also elderly men and women. They too had died. Since then he went to the city with Vimal. It was heard that Vimal also used to work in the same hospital. Vimal was also no less, she also did not talk to us. The name of the doctor was Dr. Raunak Verma.

Samayara understands that Shravan Bhogi was the one who killed Raunak and Samayara's parents. Then Samayara asks for Shravan Bhogi's number. Then Kamal tells that we do not have his number, he lives in Anjanwadi. Then Samayara asks for his photo. Kamal says to Samayara. I see that he came to my daughter's wedding. There is his photo. She sends Shravan Bhogi's photo on Samayara Rajput's WhatsApp.

After sending the photo she finds out that

The person seen in his dream is the Shravan Bhogi. Samayara gets exposed.

Samayara's dreams are not illusions, they are reality. This was the incident that happened with Samayara and fit in her mind. Even after Raunak's death, Shravan Bhogi knew that Samayara had continued the treatment-counseling of women in her hospital. He didn't like it.

Because of this, Shravan Bhogi used to steal the data of the hospital and kill women and Vimal was also involved with him.

Why is he doing this after reaching there? Why did he kill women and children? It wants to know. She vows that she herself will reach Shravan Bhogi and kill him with her own hands.

Shravan, who had dreams of Samayara, was a bhogi, but there was no proof that he was the murderer.

The police could not believe what Samayara Rajput said. She was looking somewhere in a different direction. That's why it was impossible to reach Shravan Bhogi.

Samayara was looking for such a man to reach Shravan Bhogi. The one who can catch Shravan Bhogi without caring for his life. Support Samayara. Only then Samayara remembers her college friend Vishwanath Lokhande alias Vishwa.

Vishwanath loved Samayara but Samayara did not love Vishwa. Vishwa was very good in studies and was even more expert in fighting. No college boy used to mess with Vishwa because he used to break his bones and ribs. He had proposed to Samayara but Samayara rejected him. After that he drowned himself in alcohol.

Samayara starts looking for him. Now the same ray of hope is visible. Vishwa was the only one who could catch Shravan Bhogi because Shravan Bhogi was very dangerous. Samayara finds Vishwa while searching and makes Vishwa the whole story.

In the story till now, the man who was listening to the whole story from the beginning…. It was Vishwa.

Samayara and Vishwa remember their college days.

Vishwa was full of education, understanding and worldliness. Both were also in junior college, at that time Samayara was a science student and Vishwa was studying arts. He was an expert in sociology.

Because of Vishwa's fear, no one dared to tease Samayara. But Samayara never liked Vishwa. Because she felt that she was a science topper and he was a student of Arts Sociology and Vishwa was often seen in fights. That's why she thought he was a goon. Whenever Vishwa proposed to Samayara, till then Samayara rejected and humiliated him.

Vishwa then drowned himself in alcohol but he loved only Samayara.

When Samayara asks Vishwa for help to catch Shravan Bhogi. Then Vishwa says that to catch Shravan Bhogi means to put one's hand in the mouth of death. But he will do this for Samayara. Even if he loses his life. Samayara may not love him but he will fulfill the duty of his love.

Samayara now gets thinking. She was looking at Vishwa from a different angle. But now he comes to know that Vishwa is a very nice person. Samayara shows Vishwa details of Shravan Bhogi and details study file of all the missing patients of Samayara till now.

Being a student of Vishwa Sociology, he had the practice of many practices and different levels of society. He guesses that all this is a type of malevolence. 'Narbali', Samayara Rajput also gets surprised.

Vishwa tells Samayara that we will get all the answers only after going to Shravan Bhogi's village. Both Samayara and Vishwa go to Shravan Bhogi's village. Samayara inquires about Bhogi in the village. Then they get negative comments against him. They search his house. Then Vishwa wonders that if Shravan Bhogi's wife Vimal was infatuated with him, then what was the reason that Shravan Bhogi killed Vimal? Vishwa suspects that Vimal must have something due to which she wanted to keep Shravan Bhogi under her control. And Shravan killed Vimal for the same thing.

On the other hand, Shravan Bhogi remembers Vimal at one place, he sees in the flashback that he was asking

Vimal for the proof of murder and the details of the woman. Vimal had photographs of the manner in which Dr Raunak was killed and a video was made. Vimal was threatening him with the help of that video, then Shravan Bhogi kills her and buries Vimal in the ground behind the house.

While lying at the bottom of Shravan's house, Samayara and Vishwa see a place outside the house where the ground has been dug. And many ants are seen going inside. Then Vishwa gets suspicious. When Samayara and Vishwa dig there, they find a skeleton there. There was a locket in the neck of that skeleton. Being a doctor, Samayara learns that the skeleton is of a woman. Samayara Rajput recognizes that locket as well. That locket belonged to Vimal. There was a pen drive in that locket. She knew that it was a pendrive because she had gifted that locket to Vimal. And Vimal always wore the locket around her neck. Vishwa tells Samayara that maybe Shravan Bhogi did not know this thing. Perhaps he mistook her for this locket. Shravan Bhogi considers the pen drive as a simple locket. Take out the pendrive locket from there. His suspicion turns into belief that there must be something in this pen drive against Shravan Bhogi, due to which Vimal died because of his worry.

Considering Samayara as her duty, she calls the police Pandurang Bagdi. Then Pandurang Bagdi tells that now this case has been handed over to Crime Branch Special Task Force. ACP Bhavika will handle the case. Pandurang Bagdi also tells that ACP Bhavika is finding Samayara under suspicion. Because all these women were related to

Samayara. Pandurang Bagdi sends ACP Bhavika's number to Samayara.

Samayara talks to ACP Bhavika on the phone, then ACP Bhavika immediately tells her to surrender. Samayara refuses to surrender and says that she wants to get to the bottom of it. Even if it is a case for the police, it is not a case for her because Shravan Bhogi had killed Samayara's parents and good friend Raunak. Samayara also tells that Shravan Bhogi had killed Vimal and buried him in the village and she should send the local police there. Vishwa and Samayara leave from there. ACP Bhavika sends the local police there. Here the police find 8 female patients who were taking treatment from Dr. Samayara and have gone to their respective villages. Police issues warrant against Samayara Rajput. Samayara's assistant Vicky calls her and tells her that a warrant has been issued against her.

The police started chasing Samayara. Both Samayara and Vishwas run away. The police learn that Samayara's friend, Vishwa is with her. Bhavika also interrogates both Vicky and Jenny. The police issue warrants against both Samayara and Vishwa.

Vishwa comes to know that this bloodshed is happening only for human sacrifice and Vishwa comes to the logic that it was happening because Shravan Bhogi could not become a father due to the accident. That's why maybe he is giving this male sacrifice to become a father. And as described in the Narbali Ramayana story, King Ahiravan of Patal Lok, the brother of Dashanand Ravana, used to give it to his goddess Kamna. Shravan Bhogi used to offer this human sacrifice in every full moon night.

After searching in the library and cyber related to Kamna Devi and Ahiravan, Vishwa comes to know that Shravan Bhogi will kill 11 people. And if we see the method of drawing blood in it, then it was orbiting on the map of India. He was choosing a deserted temple every time to kill him. Blood was being shed around the deserted temple.

Samayara and Vishwa go to Bulandpur. Vishwa asks Samayara if she has any patient in Bulandpur? And one patient, one of her patients used to live in Bulandpur. They goes there to reach that patient, but before that she disappears. And her dead body is also found. Shravan Bhogi kills her in Bulandpur. Samayara then calls Bhavika and informs her. Help them, we should help in catching Shravan Bhogi as soon as possible. Bhavika warns to surrender immediately. Samayara told that now is not the time to surrender because after 5 days there is going to be another bloodshed on Poornima and this time he is going to pick up someone from Madhya Pradesh. Both leave from there.

Police also reaches Chittorgarh. Samayara's patient goes for a walk in the market. Then Shravan kidnaps her. Then the police comes there to save her. But after killing many policemen, Shravan takes her away. Samayara Rajput and Vishwas both also come there but both of them are unable to save themselves. Bhavika holds Samayara and Vishwa at gunpoint. Then both of them request her that both of them will help to reach him and will also hand over themselves over to the law. Because after 3 days it is Amavas and it was not known where Shravan would take the patient. both run away

Both of them also do a lot of research. But it is not known anywhere. Jenny and Vicky also come to help Samayara. Vishwa finds out that there is Patal Lok in Chittorgarh, such a world is believed and Ahiravan used to sacrifice himself in Patal Lok. So Vishwa has a doubt that whether this happens or not, he will fulfill his vow by giving 11 Bali in Patal Lok itself. He comes to know that there is an old temple of Kamna Devi in an old Patalalok area in the old forests above and no one goes there in the forest. Both Samayara and Vishwas go to the forest. The police also come to know about that forest while chasing. It has been seen Samayara and vishwa in Pithoragarh, this news is also available.

Officer Bhavika also enters the forest with some police. Shravan Bhogi kills some police officers. He sets fire to the forest, because of which all the other policemen retreat. There, he starts worshiping Kamna Devi and goes to sacrifice the woman when both Samayara and Vishwas stop him. There is a huge scuffle between the two. Shravan Bhogi attacks Jenny and Vicky. Samayara Rajput stabs in Shravan Bogi's stomach and tears open his stomach. Vishwa and Bhavika also get injured. Samayara expresses her love for Vishwa, both of them also laugh.

Story No.03 Ghera

This story is from a small town. Rajeev Adarkar and Kiara Adarkar. This family couple lives in a residential complex. Both of them have been married for 5 years. Both are also career ambitious. Both have to touch heights in their respective careers. Rajiv runs an ad agency business, while Kiara is a successful blogger. Rajiv's business is going well. Kiara has crores of followers, who read her blog with interest and ask her for solutions to their problems.

The family routine of both was fixed in the same way. Get up early in the morning. To have breakfast in the house, Fulwari, the maidservant, goes to the house after doing the morning meal, clothes, utensils and cleaning. Kiara is busy preparing for her husband's departure. Rajeev also used to leave for office in a hurry with his laptop. He used to take homemade food only while going to work. There is a lot of love between both husband and wife. Kiara has started her blogging after Rajeev's departure. After that the maid Fulwari goes away after completing the work. Kiara starts her routine by calling her parents and at the same time praises her husband Rajeev. Rajeev used to work in his agency office throughout the day and Kiara used to visit the market every evening to buy household items and vegetables. Rajeev also used to come home in the evening, both used to eat food with great love and used to go to sleep at night chatting.

Another family lived in the same complex. Prayag Saxena and Manva Saxena, this couple used to live in another house. Both also used to do private jobs. Both are also very happy with their household. Both Prayag and Manwa were a happy couple. It had been 2 years since their marriage. Both also loved each other very much. It is believed that they are made for each other.

The next day, like the daily routine, Rajeev goes to his office and Kiara gets busy with her work. The maidservant also leaves after finishing her work. In the afternoon Manva dials a phone number from the mobile and asks "Have you eaten or not and what are you doing?" On the other hand, Rajeev is talking from his office "Tumhe miss kar raha tha sweet heart" Both are talking on the phone call of friendbook. The friend book was a social site, through which we could call and chat online. It is clearly known from the words of both that both of them were having an affair. After completing her blogging work, Kiara again goes to the market for shopping in the evening and brings the necessary items. Rajeev leaves for home and while coming in the car keeps on chatting continuously with Manwa on friend book. After coming home, Rajeev and Kiara laugh looking at each other. Rajiv also expresses his happiness. On the other hand Manwa also smiles seeing Prayag. Both the couples first eat a little food, discuss daily routine and show the common bonding of the couple. Talk lovingly to each other. fall asleep.

The next day Manwa talks on the phone that "I will not show my identity to you outside and do not even try to talk to me. I do not want that because of this the future of both of us should be spoiled. Even if you try to come near I will also not reveal my identity." On the other hand, Rajeev tells on the phone, "I love you." Then Manwa says in front that "If you want to marry me, then you will have to leave your wife Kiara. First you will divorce her, later I will divorce my husband and I do not want to talk much about this." "

Rajeev could not leave Kiara so easily by divorcing her. Because the flat in which Rajeev lived belonged to Kiara. Kiara was also a partner in Rajeev's ad agency. Rajeev would have been in loss if Kiara was given a divorce. On top of that, Kiara's nature was very aggressive. She could not tolerate this affair of Rajeev. That's why Rajiv went under a lot of tension. He started thinking how to break the marriage relationship with Kiara.

Chatting used to go on in both of them often on friend book. Manva tells Rajeev that if he wants to have a relationship with her, then he will have to leave Kiara. Rajiv tells her about his compulsion that he cannot give Kiara a divorce. Manwa tells him in chatting that don't talk to her.

Rajeev gets thinking how to find a solution in this? It takes a lot of thinking on this matter. Also finds solutions on the Internet. Finally, he decides that he will kill Kiara in such a way that the other person will get an accident. Because of this no doubt will fall on him and everything will be his after Kiara's death. Thinking all this, he informs Manwa about this. But Manva tells her that its okay then finish it and call me only later as she cannot trust him like that. Manva tells from the other side, "As soon as the work is over, tell me immediately. I will be ready. We have to take this risk because it is very important for our future."

In these circumstances, Rajeev goes under tension. Rajeev's health deteriorates due to tension. He starts feeling dizzy. Kiara takes Rajeev to the doctor. The doctor asks to test Rajeev and he is asked for bed rest. Rajiv becomes weak. During this, Kiara takes care of Rajeev very well. His report is bad, then Kiara alerts him to rest and not go to work. Kiara takes good care of him by giving him regular pills.

Something else was going on in Rajeev's mind. He was thinking that in the meantime I will kill Kiara in any way. He removes the heater wire in the bathroom as Kiara gets fond of touching it and tries to kill her by taking out the gas pipe. But both times, due to this thing being noticed by Fulwari, Kiara is saved both times. On the other hand, Manwa and her husband Prayag seem happy in their household. Laughing and playing are seen in love.

Rajeev plans to kill Kiara with certainty this time. Kiara had asthma, he intentionally blows dust near her. Because of that Kiara gets an asthmatic attack and she uses her inhaler. Rajiv had kept her inhaler hidden. He closes the door. Kiara suffers because of the attack. Rajeev closes the door and starts listening to her agony. After the sound stops, he slowly opens the door. Then Kiara's mouth is open. Rajiv becomes happy that Kiara is dead and immediately calls Manva. But he hears the ring of the phone in the same room. When he looks around, he sees Kiara's phone ringing. When Rajeev disconnects the phone and dials again, only Kiara's phone rings. He calls three to four times, but Kiara's phone rings again and again. One gets thinking that I am calling Manwa but why is Kiara's phone ringing? Rajeev opens the mobile phone by touching Kiara's thumb. From that in the friendbook, he comes to know that it was Kiara who was talking and chatting with him after making her profile. He gets tense. Rajeev drinks water looking at Kiara and eats her pills in tension. After a while he gets paralysis attack.

On the other hand Manwa used to talk to her husband every time. The reason for this was that Manwa and her husband Prayag used to work in the same private limited company. According to the rules of that company, no husband and wife could work in the same company. That's why both of them used to hide their identity from each other and used to talk silently on the phone. Prayag was about to change his company soon. Manva had no affair with Rajeev. She didn't even know him well. Both of them knew that Rajeev Kiara lives in their own complex.

Here Kiara opens her eyes and gets up. Kiara had another emergency inhaler in the bed room. Seeing her husband Rajiv lying paralyzed, she starts laughing. Rajeev understood that, Kiara knew what he was thinking about Manva and her. Kiara tells him that she knew that Rajeev wanted Manva. But she wanted to know more what he was thinking about Kiara. Then Kiara creates a fake account in the name of Manwa and starts chatting and calling Rajeev from it. When Rajeev tells about his plan to kill Kiara as Manva, she gets angry. Because Rajeev was cheating on her love. When Rajeev was ill, Kiara changes his reports. Also gives him wrong medicines. Gives such food that it is food poison.

 Rajeev is hospitalised. The doctor tells Kiara that "he has got an infection because of taking the wrong pills. Because of that he is paralyzed." Rajeev had heavy sugar and due to taking wrong pills, he had a reaction and became paralyzed. Along with this, he also had a food infection. Kiara brings wrong pills which Rajeev was consuming continuously. Kiara used to give him food which is not good for his health for food poisoning. Rajeev was unable to speak due to being paralyzed. Doctor says "He will take a lot of time to recover. It would be better if you treat him at home. Shift to home so you can take care of him there and appoint a nurse". After the doctor leaves, she says to Rajeev, "Paralyzing you was half my revenge. I loved you and you wanted to kill me in return for property and another woman? Let's go home now and then I will decide how to treat you." Because your death should look like an accident or natural to the world." Kiara laughs out loud. Rajeev cries in agony.

Story No.04 Dandak

There was a very forested tehsil area named Shikanji-Madhopur. There were many small villages and settlements. The head of the village was rich and the rest of the settlements were very poor. Because of poverty there were many thieves and dacoits are increased. Poverty was touching its peak. One such night, a man and a woman go out in search of someone in a deserted area. He was looking for someone. Suddenly they are attacked. The person who attacks is like a witch. Whose hair remains big. The witch man dressed in strange clothes kills the woman. Chudel injures that person and pelts his head with a stone. The witch takes the woman's blood in her bowl and mixes some things after looking at the book. In the end she stuffs a key in a silver box and laughs out loud.

A poor family lived in a township somewhere far away from this incident.

Both Ghisu and Champa were husband and wife. Champa was very hardworking. She used to go to people's fields and work as a labourer to run the house. On the other hand, Ghisu was poor, but he did not appreciate the hard work of his wife. He was very lazy. He knew how to drink alcohol and roam here and there on the pretext of work. Champa loved him very much. She used to convince him that he should do some work, so that some money could be collected. But Ghisu was greedy. He needed money immediately, that too without any effort. Both of them were not educated, but they were intelligent. They knew very well how to face difficult situations. Their 3 children died due to illness. Ghisu and Champa did not have money for their treatment. Since then Ghisu drowned himself in alcohol. Due to poverty and lack of money, they used to be frequent quarrels between them. Ghisu used to commit thefts in the village, sell the stolen goods outside and collect the money. On the other hand, Champa fasts and prays to God. Champa asked for various wishes, but to no avail. Neither did their poverty go away, nor did they have a child, nor did Ghisu give up his habit of drinking and stealing. Some days go by like this....

There was an old tree in the colony. Under which the deities were established. Which was also called the Pedwala temple of Basti. Ghisu and Champa go there. Champa asks God for a vow and says, "If you make us happy, then Ghisu will do hundred parikramas of the temple mountain." Ghisu gets angry over this. He gets into a fight with Champa. Ghisu hits Champa, then Champa pushes him away. Ghisu falls on the ground, then he realizes that he saw something in the bushes. There he sees a small silver box. Every now and then someone is watching them from behind the bushes. He lifts the box. Champa also becomes happy. Both of them feel that there must be some treasure in it. They both hide the small box. Let's look around to see if anyone has seen them. Someone is watching them from behind the bushes. He hides the box and brings it to his home. When the chest is opened slowly in the house, then they see a key inside and a map on a cloth. This complete symbolic map remains. But both of them come to know that this is the key to some treasure and this is the map to reach that treasure. But how to reach? They don't know this. Both of them keep thinking in the same. Had a map in hand. But cannot read the language there.

Later both of them come to know that from where the beginning of the map begins. There is a temple of the old tree god of the village. There both of them see the idols of God marked on the map. On the next morning, Ghisu sets out in search of the way after looking at the map thoroughly. If he finds the way after walking for 3 hours, then he comes back from there. Ghisu tells Champa that he has found the way. That path turns into 3 paths while walking from the temple, roaming around. One of those paths is deserted, which leads to the ancient crematorium. Where no one goes and there is a hilly area. The area of the ancient cemetery is completely deserted. Champa also comes to know by looking at the map that at the end there is a sign of moon and full moon means full moon. In the end, the pedestal and the big chest are visible. Full moon remains after 2 days. They both set out to find the treasure on the same day. After leaving the forest in the afternoon, they stay there till evening. After nightfall, with a torch, they go to the ancient crematorium area in search of a pedestal and a big chest in the forest.

Here he sees a chair and a chest next to it. They both become happy. As soon as he opens the box, he sees a dead body inside. Champa shouts. When both of them try to run away from there, the former witch attacks them. They also compete with the witch to escape. During this, the witch kills Champa. Ghisu also retaliates to the witch's attack. Ghisu goes to Champa's dead body crying and tries to wake her up, starts crying, then the witch comes from behind and hits him on the head with a stick. Chudel takes Champa's blood in her bowl and mixes some things in it after seeing the book. In the end, she puts a key in a silver box. Chudel that blood is applied on that dried corpse.

After Ghisu regains consciousness, the witch tells him that he is not a witch but a man and the dried-up dead body in this box is the dead body of his wife. He tells the injured Ghisu.

That, "My wife and I had found this chest and the key. Seeing the map, we both came here in greed of treasure, then a man like me was already sitting here. He attacked both of us and killed my wife. And when he opened this big box, there was the dead body of that man's wife inside it. He put my wife's blood on it and took it away. He told me that you also put your wife's dead body in this box and this key in the forest. Leave it. Some husband and wife will come to take this box, then kill his wife and apply her blood on your wife's dead body, then your wife will be alive in 24 hours. This is what I did in picture form inside this book. It is a complete ritual. You also leave this key in the forest, then husband and wife will come to take this box, then after killing his wife, apply her blood on your wife's dead body, then your wife will be alive in 24 hours. ."

Ghisu starts crying. The chudel man is carrying away the dead body of his wife when Ghisu kills him with a stone. Then Ghisu keeps his wife's dead body in the chest thinking something while crying. He wears the witch man's clothes himself, then sits on a chair and starts crying. After a few days in the morning, a person sees that small silver box near the tree temple.

Story No.05 : Question Mark

Story No.05 Question Mark

A dark place…..a slow beat in a dark place a phone rings…..an alarm goes off in the phone. The phone is ringing like an alarm. But no one picks up. Suddenly a person is seen sleeping beside the phone. This person is a young man in his thirties, who is fast asleep. Gets up from sleep and takes the phone. Alarm Buzzer is the alarm going off, he turns it off and goes back to sleep. After five minutes there is an alarm again. Now the person looks at the time and turns off and goes back to sleep.

The person opens his eyes again. When he opens his eyes, he thinks it's pitch black around him, so he gropes around with his hands. He realizes that I am in a box and starts thinking. How do I close this box? How did you come to this? How am I here? While being bombarded with this question, he feels…. He starts trying to get out.. How am I here? As he continues to be bombarded with this question, he realizes….. I… I… mean? who He starts remembering his own name… Shardul Kale…… He was an IT expert freelancer. His wife Shama Kale was a counsellor.

He shouts angrily and kicks the box. That box is like a grave and he wants to get out of it. He finds chewing gum in his pocket. He also has a key in his pocket, close by, he tries to open the box using it, but stops when dirt starts falling on his face, trying to get out he slowly falls asleep.

After a short while after falling asleep, he falls into a sound sleep. He gets up again. When he wakes up again, he easily picks up the alarmed mobile phone next to him. Checks the mobile phone. Dials his known number but the outgoing service is off. He is shocked. He can easily press any number. But the outgoing service is off. No number is required. Check out its inner gallery and other features. He then looks at the phone. So there is only one number in the phone book and that number is written here as Call Me.

He dials the number in the phone book, but on the other hand, he hears a sleeping girl in the dark and she is fast asleep and does not pick up the phone. Now calls again. Now slowly she wakes up and she picks up the phone with sleep on her eyes she starts talking on the phone in deep sleep. So he calls her and asks. "I'm stuck in a box. Save me please save me" she asks his name "where is he stuck?" Asks "Who are you? What is your background?" Sara asks but like that Shardul tells her his name where does he live...but where is he stuck? He cannot tell which part he is in. So in her sleep she thinks someone is joking so she hangs up the phone and goes to sleep.

He calls her again and asks for help again, on which she angrily says, "Hey, you don't tell me where you are stuck... How can I help you?" After that she hangs up the phone again and he calls again to get help and after calling he asks, she says "You should just dial 100 and call the police. But he says that "I am not receiving any number from my phone. No need for another number... Your number was saved and your number was dialed. Please call from your number and call the police please…. At least contact my wife..her number is 404084422 Shama Kale." But she says "Ok call her and tell her your husband is stuck in the box...Ok?"

She hangs up again. And as she looks around, she notices that there are so many boxes, and just as he described that he is closed in a box, there are also descriptions around that she is also closed in a box, and she struggles to get out. She cries and becomes lonely and desperate.

As Shardul calls her again, she tells him that she too is locked in a box.

He asks her her name

She says Rasika Bhalerao ... Husband .. Shekhar Bhalerao ... Family Business Global Tours and Travel ....

Shardul asks to call his wife from her phone.

She calls Shekhar but doesn't pick up... After a while Shardul's wife calls Shama. The ring rings. And the sound comes from there. Rasika tells her "Your husband is stuck in the box." Shardul's wife says how is it possible Shardul is at home. As she informs Shardul about this, he tells Rasika to call Shama again, but this time Shama gets angry and threatens to call the police, after which she gets a reply that her number has been blocked.

Rasika calls her husband i.e. Shekhar when she doesn't pick up the number, she calls Shardul and tells about his wife... He gets shocked and she doubts whether Shardul is telling the truth or lying. Rasika gives Shardul Shekhar's number and asks him to call Shekhar, Shekhar also replies that Rasika is at home. He informs Rasika about this.

As he starts to look around, he notices a photo in his pocket. There is a young woman with him in that photo, he wonders. He remembers the girl and the night he spent with her in Bangkok.

A year ago in Bangkok, he also saw the young woman suddenly in a hotel, neither of them had asked each other's name. Who put this photo in his, his, pocket in such a condition today after a year? He started thinking about it. The young woman in that photo was none other than Rasika Bhalerao who was confined in another box. Shardul knew her only by face.

Rasika looks around and also sees a photo beside her. In her own pocket… This is the same photo that Shardul has where the young man in the photo sees and remembers her in Bangkok. Wondering who took this photo? And after a year this incident happened and knowingly delivered this photo to her?

She calls him again and informs him regarding the call and talk… Both of them talk on the phone and while talking to each other they get a sense that someone has deliberately tied them here and they both describe to each other what they look like. They have a relationship with each other but they do not know each other by name.

………As both of them tell each other about the photo, they realize that they are the only two who met in Bangkok and that someone has locked them here. Both of them start talking on the phone and after that both of them keep the phone and start looking for a way to escape.

When he feels something near his feet, he looks around with a light, not a phone, then he sees that there is a hole in the corner of the box for oxygen to come in……. He is sure to get that hole oxygen. He also informs her about that.

He notices something near his feet. He slowly pulls the object up with his feet. So he sees that it is a kind of time bomb. Its timer has started slowly. He also calls her.. She also has something near her feet, she pulls it up and there is a similar bomb in her hand, the timer is also on.

Both their hearts start racing as the time bomb timer goes off… time is running out… time is approaching…… both think about it and how to stop the time?

As time goes by…. As such, the lives of both of them start to turn upside down.

Suddenly there is a video call on both of their phones. It is a conference call in which both Shardul's wife Shama and Rasika's husband Shekhar are present. Both of them reminice about how they were cheated by both and how they tried to kill them time and time again to grab their property.

They also tell both of them that there is a red button on the back of both of their bombs. Pressing this button will stop the bomb timer and the bomb will not explode. But if shardul presses the red button then rasika's bomb will explode instantly and if rasika presses the button then rasika's bomb will stop timing… shardul's bomb will explode instantly and the phone keeps going…..

There is still one minute left for the bomb to go off. Both of them stare at the red button… both of them get ready to push the button. Shekhar and Shama both hear the sound of a bomb going off… Both turn and find Shardul and Rasika's photo behind them…. They leave with a flower in front of the photo….
